This Book Belongs to:

Thank You for Your Purchase!

Enjoyed your Ruby Doodle coloring book? We'd love to hear from you! Write a review and share your experience with others. Your feedback helps us create even better books. Thank you!

★ ★ ★ ★ ★

Unlock Your Bonus Coloring Pages! Scan this QR code to receive free pages from the Ruby Doodle collection. Discover new designs and expand your creative journey. Happy coloring!